Abel Hernández-Muñoz
José Blas Pérez-Silva
José M. Ramos

THE SPIDERS (ARACHNIDA: ARANEAE) OF THE GUAMUHAYA MOUNTAINS, CUBA

Abel Hernández-Muñoz
José Blas Pérez-Silva
José M. Ramos

THE SPIDERS (ARACHNIDA: ARANEAE) OF THE GUAMUHAYA MOUNTAINS, CUBA

THE GUAMUHAYA MOUNTAIN SPIDERS.

ScienciaScripts

Imprint

Any brand names and product names mentioned in this book are subject to trademark, brand or patent protection and are trademarks or registered trademarks of their respective holders. The use of brand names, product names, common names, trade names, product descriptions etc. even without a particular marking in this work is in no way to be construed to mean that such names may be regarded as unrestricted in respect of trademark and brand protection legislation and could thus be used by anyone.

Cover image: www.ingimage.com

This book is a translation from the original published under ISBN 978-620-3-03355-7.

Publisher:
Sciencia Scripts
is a trademark of
International Book Market Service Ltd., member of OmniScriptum Publishing Group
17 Meldrum Street, Beau Bassin 71504, Mauritius
Printed at: see last page
ISBN: 978-620-3-18781-6

SPIDERS (ARTRHOPODA: ARACHNIDA: ARANEAE) OF THE GUAMUHAYA MOUNTAINS, CUBA.

Authors: Abel Hernández Muñoz*, José Blas Pérez Silva**, José Manuel Ramos Hernández***, Osmany Manuel Rodríguez Pino****, Alexei Rodríguez Lorenzo****

Sancti Spíritus,

Cuba 2020

THE SPIDERS (ARTRHOPODA: ARACHNIDA: ARANEAE) OF THE GUAMUHAYA MOUNTAINS, CUBA

Authors: Abel Hernández Muñoz*, José Blas Pérez Silva**, José Manuel Ramos Hernández***, Osmany Manuel Rodríguez Pino****, Alexei Rodríguez Lorenzo****

Affiliation: *Natural History Museum of Sancti Spíritus "Juan Gundlach"; **Center of Environmental Services, Sancti Spíritus; ***Department of Vegetal Health of the Provincial Delegation of MINAGRI; "Samá" Group of the Speleological Society of Cuba.

SUMMARY: The spider fauna that lives in the Mountains of Guamuhaya, south-central region of the island of Cuba, is composed of 239 species belonging to eight orders and 55 families, being Araneae (194 species, 40 families) and Opiliones (20 species, seven families) the most diverse. Among the most interesting elements, 66 local endemisms stand out, some of them restricted to only one cave. The spider *Gaucelmus augustinus* Keyserling, 1884 (Nesticidae) constitutes a new record for Cuba's fauna. Numerous species are registered here, for the first time, for Guamuhaya.

Keywords: Spiders, fauna, Antilles, Cuba.

THE SPIDERS (ARTRHOPODA: ARACHNIDA: ARANEAE) OF THE GUAMUHAYA MOUNTAINS, CUBA

ABSTRACT. The recorded spiders from Guamuhaya massif in central Cuba belong to 239 species of 55 families, and eight orders. Araneae (194 species, 40 families) and Opiliones (20 species, seven families) are the most diverse. Sixty six of these species are local endemism, some of them inhabiting a single cave. Gaucelmus augustinus Keyserling, 1884 (Nesticidae) is herein recorded for the first time from Cuba. Several species are herein recorded, for the first time, for Guamuhaya.

Key words: Spiders, faunistic, West Indies, Cuba.

INDEX

INTRODUCTION

In the south of Cuba's central region is one of the country's main mountainous areas: the Guamuhaya Mountains, often referred to as El Escambray (fig. 1). With an extension of 80 km and oriented from east to west, this orographic system constitutes an important center of evolution of Cuban biota (Samek, 1973; Iturralde-Vinent, 1988). From the political-administrative point of view, it is distributed among the provinces of Villa Clara (northwest), Sancti Spíritus (almost all the eastern half) and Cienfuegos (great part of the western half).

In general terms, the Mountains are divided into two mountainous groups, separated by the depression of Agabama River: the most western group is known as "Alturas de Trinidad" (sometimes referred to in literature as "Mountains of Trinidad" or "Sierras de Trinidad") and contains the maximum heights of central Cuba: San Juan Peak (1140 masl) and Potrerillo Peak (931 masl). The eastern part constitutes the Heights (or Mountain Range) of Sancti Spíritus and its maximum elevation is in the Banao Hill (842 msnm). According to Barranco & Diaz (1989), the massif has a humid tropical climate with rain all year long and a warm temperate climate with rain all year long, according to the modified classification of Köppen.

In the foothills the semi-deciduous forest predominates, in the karst areas the vegetation is mogote and mountain rainforest on schist; while towards the higher parts there is also the evergreen forest. In addition, there are plantations of male pine (Pinus caribaea) and coffee plantations. The average annual rainfall in the area reaches 2000 mm.

In general terms, the arachnophauna of the Guamuhaya Mountains and its foothills has been poorly studied. During the first four decades of the 20th century, explorations were restricted almost exclusively to the westernmost part, in areas of the "Soledad" sugar mill (now "Pepito Tey") and the adjacent Botanical Garden. However, it is evident that "Soledad", as it appears in the data collected from the specimens, is almost always a simple reference point, since this was the best known geographical site. In many cases, the materials came from the nearby foothills of Guamuhaya, which at that time were much more heavily forested than they are today. In fact, some localities consigned as "Soledad mountains",

"Soledad, San José sumit" and "Soledad, Trinidad Mountains" (Bryant, 1940), suggest so.

From the second half of the twentieth century, the attention of the arachnologists is directed to Topes de Collantes, due to the facilities of work and road communication, because the facilities of the forest station located there and later the Faculty of Mountain of the Central University of Las Villas, provided important logistical support.

One location that also benefited from the communication and logistical facilities was the city of Trinidad. In its surroundings, which constitute the confluence point of the Trinidad Heights and the coastal plain, many naturalists and visitors made sporadic incursions and took some samples of its arachnophauna.

Thus, it should not be surprising that these three locations (Soledad, Topes de Collantes and Trinidad) are the ones that appear registered with the largest number of taxa.

Regarding the study of the araneofauna that populates this mountainous massif and its foothills, Bryant's work (1940) deserves a special mention, in which a considerable number of species were described or registered for the area. This monograph, one of the most important ones made so far about Cuban araneofauna, was almost exclusively based on the collections made by North American students and naturalists who visited or explored this country between 1900 and 1938, being the prestigious entomologist P. J. Darlington one of the most relevant collectors.

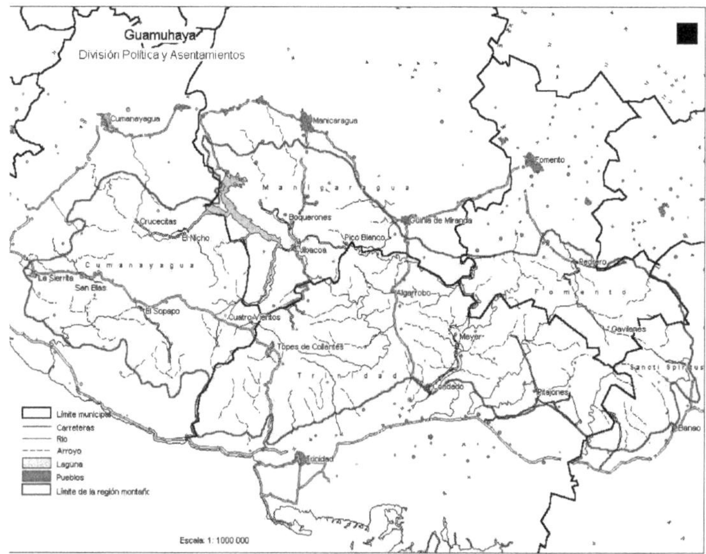

Map 1. Location of the study area (Guamuhaya) of the most relevant collectors.

The only list published about the spiders of Guamuhaya is that of Alayón (1994), although the precise locations were not indicated, but only "Macizo de Guamuhaya". As for the cave arachnophauna, Ramos & Hernández (2007) registered 26 species belonging to six orders (including a parasitic bat tick), found in eight caves in the area of Aguacate, municipality of Cumanayagua, province of Cienfuegos, although only a quarter of them were identified to the species level. Although this last work represents an important contribution to the biospeleology of this orographic system, its dissemination has been very limited, since it was published in digital format (CD-R).

This contribution compiles all available information on the taxonomic composition and geographical distribution of the Guamuhaya Mountain Spiders, while providing new records of taxa and locations, as well as natural history data of some species. Some inaccuracies detected in the bibliography are also clarified.

STUDY AREA

Geographical location

Guamuhaya is a mountainous system located in the provinces of Sancti Spíritus, Villa Clara and Cienfuegos, in the central zone of Cuba. Its highest point is Pico San Juan which reaches 1140 meters above sea level. At its feet is the city of Trinidad. It is the third most important mountainous system in the island of Cuba, after the Sierra Maestra and the Guaniguanico mountain system. The mountain range received the name of Guamuhaya from the natives that inhabited the area before the Europeans arrived in 1492.

The Guamuhaya Massif is subdivided into two large groups, Alturas de Trinidad to the Northwest and Alturas de Sancti Spíritus to the Southeast. It is characterized by its abrupt ravines and deep valleys, the exuberance of its vegetation, the great diversity of its native flora and fauna, great systems of caves, beautiful landscapes, and its rivers, waterfalls and crystalline water pots. The territory is rich in mineral resources but they are not being exploited due to the danger that their exploitation represents to the environment.

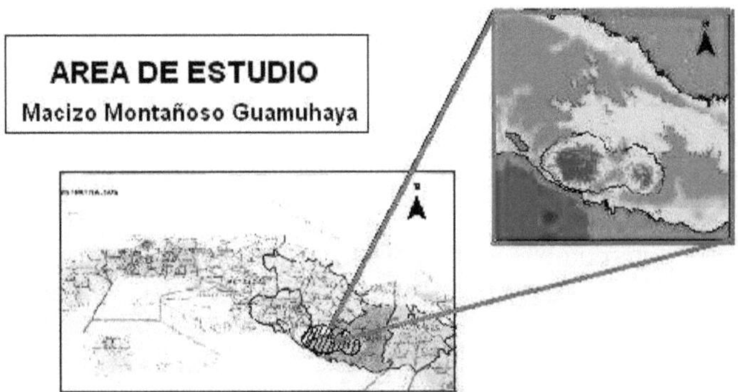

Figure 1. Study area.

It is located in the south of the central portion of Cuba. It occupies a surface of 1948 km² (approximately 11% of the mountainous area of Cuba). From the political-administrative point of view, the region occupies parts of the provinces of Sancti Spíritus (42%), Villa Clara (33%), Cienfuegos (15%). It limits to the North with the heights of Santa Clara, to the East with the valley of the Zaza river, to the West with the valley of the Arimao river and to the South with the Caribbean Sea.

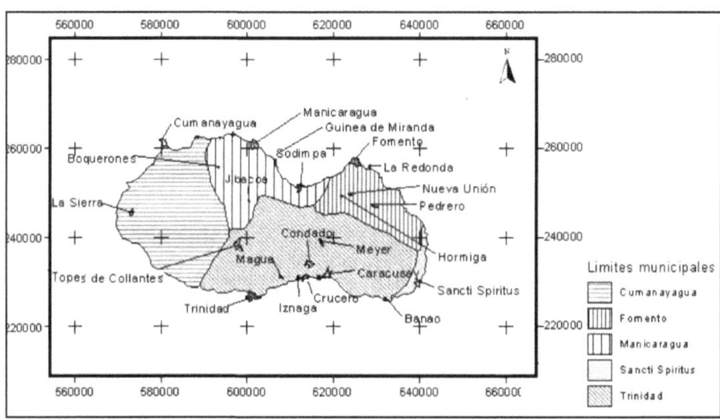

Map 2. Municipalities of the mountainous massif of Guamuhaya. Source: Caves, *et al.* (2006)

Geological Structure

The main rocks found in the region are metamorphic, considered the oldest in the country. The mountains of Guamuhaya generally present a folding or arched structure.

In the south of Central Cuba, in the area comprised by Trinidad anticlinarium, there are different lithological complexes corresponding to different geotectonic environments, so we have the so-called metamorphic complex or Escambray Massif, whose protoliths essentially correspond to a passive continental margin, the Mabujina amphibolytic complex, the sedimentary vulcanogenic complex of the Cretaceous arch and the Manicaragua granite complex.

The last three correspond to the geotectonic environment of a volcanic arc with its foundation of officious character. Finally we have the sedimentary cover constituted by young accumulations, transgressively deposited on the previous complexes.

<u>Escambray Complex</u>

It is mainly constituted by Jurassic and Cretaceous sequences deposited in a great extent in a passive continental margin. The sequences from the Lower Jurassic to the basal part of the Upper Jurassic, are of a terrestrial character with intercalations of limestone, silicite and basic volcanic rocks that occasionally form bodies of up to tens of meters of power.

The rocks from the Upper Jurassic to the Lower Cretaceous are essentially carbonate, while the upper part of the cut to the Upper Cretaceous has a heterogeneous protolith, highlighting calcareous rocks, silicites, clay, terriens and basic vulcanites that form powerful bodies in some parts of the cut. These sequences appear forming nappes and tectonic scales of different orders and generations, manifesting differences in the characteristics of rocky cut in the units of main nappes (Millan and Somin, 1985 a, b), (Millan, 1990, 1992 a).

Within this Massif there are small bodies of antigoritic serpentine, metagabre and metadiobases, sometimes intimately associated, which suggest fragments of metamorphosed ophthalmic rocks, limiting different tectonic mantles and scales, highlighting even serpentine melanges in some cases.

A larger body whose protolith is mainly a toleic basalt, was converted into high pressure amphibolites (Yayabo Fm.). These rocks can also be intraformational, constituting bodies located within Jurassic and Cretaceous formations, both in the metamorphosed ones under high pressure conditions, and in the facies of green shales (Millan, 1992 a).

The Escambray's sequences stand out because of the effects of two metamorphic events, the first one is a high pressure metamorphism that affected a great part of the Jurassic rocks and different officious bodies. The second later event corresponds to the facie of the green shales metamorphosing the Cretaceous, part of the Jurassic as well as other ophthalmic bodies. The latter partially diaphthorized the metamorphosed rocks under high pressure conditions. The metamorphic zoning established in both events shows an inverted disposition with respect to the anti-formic structure of the Massif (Millan, 1992a).

The rocks of the Escambray were affected by different stages of folding during its metamorphism up to 5 or 6 phases, the oldest phase has a transverse tendency highlighting folds of similar style until isoclinals, which can reach hundreds of meters and are associated with a metamorphic schistosity.

The second phase of folding is the most important, relating to similar folds up to isoclinals that reach hundreds of meters and several kilometers. It has a metamorphic schistosity and a tectonic alignment oriented towards the Northwest, which is evident in almost all the cuts of the Massif. During this stage, the two anti-formic megastructures that played the main role in the final conformation of the massif were formed (Millan, 1990).

Relief

The altitude of the area varies between 700 and 900 m, with some areas above 1000 m, among which the Pico San Juan stands out with 1140 m. It forms mountain and premountain chains with sharp crests and steep slopes, with very intense erosive processes; the valleys are narrow, with small alluvial plains.

The relief responds directly to the lithology and structural characteristics of the mountain system. On metamorphic schists, the drainage network is deeply embedded, forming mountain and premountain chains with sharp crests and steep slopes, with very intense erosive and denudative processes, often associated with gravitational phenomena.

The valleys are narrow and form small, flat-bottomed, generally dissected alluvial plains. To the south of the mountain system, especially towards the Trinidad area, a series of abrasive, carious terraces can be observed, gently sloping towards the sea and dissected by the drainage network (Zamora, 2016).

Figure 2. Relief of Guamuhaya.

Many of the elevations of the Trinidad group show the shape of cones like the mogotes of the Sierra de los Órganos, with caves drilled into their almost vertical slopes. At the same time, some open valleys in the orographic massif do not offer superficial drainage, but rather subway, through caves. Some drainage of these basins, almost closed, are made through narrow canyons and waterfalls as was the case of the Habanilla River.

Trinidad Mountain Range is higher than Sancti Spíritus; in the first one, the region of Topes de Collantes stands out, with Potrerillo Peak, at 931 meters above sea level, only surpassed in this mountain range by San Juan Peak, the highest elevation in the central Cuban region, at 1140 m; in the second one, Banao Heights stand out, at 842 m. The mountains of Guamuhaya Massif constitute one of the morphostructural singularities of Cuban archipelago, corresponding to neotectonic uprisings (upper Miocene - Quaternary), in the form of dome, with a complex internal structure and concentric disposition of its altitudinal steps, that conserves the tendency of the paleogenic ascents (Portela et.al, 1989). Investigations carried out by this same author, revealed the existence of deluvial deposits in the upper summit levels, as in La Coca, derived from the washing of higher surfaces, probably older than the current ones.

This mountainous system is located to the South of the central portion of the Island of Cuba, limiting to the North with Santa Clara Heights; to the East and West, with the valleys of Zaza and Arimao Rivers respectively; and to the South, with Jagua Trench, in the Caribbean Sea. They have an extension of 1948 Km2 (approximately 11% of the mountainous area of Cuba) and have average heights between 700 and 900 m, with some heights that exceed 1000 m, among which San Juan Peak (1140 m) stands out, as the highest point of the massif.

From the morphological-structural point of view, this great orographic unit is subdivided into two large blocks: the Trinity Mountains and the Sancti Spíritus Mountains, separated from each other by the engraving type depression of the Agabama River.

According to the world hypsometric classification of mountains, Guamuhaya is a low altitude mountain system, very dissected, where a relief of great complexity predominates, determined by the high degree of metamorphism; formed by slopes with an average slope greater than 18º and values of density of dismemberment of the drainage network greater than 2 km/km² for horizontal dissection and about 300 m deep for vertical dissection.

The dome - block configuration of the Guamuhaya mountain system has been developed, during the neotectonic stage, on the folded base of the Jurassic methyl and metacarbonate complexes. The neotectonic ascent of the system is directly reflected in the relief through the quasi-concentric arrangement or the levelling surfaces and the altitudinal floors, the ring design of the fluvial network, the spatial distribution of the age of the relief, the arrangement of the tectonic scarps that limit it and other indicators.

On the other hand, the elements of its ancient structure, are passively reflected in the relief, in the form of mono-clinical limestone hats and armored peaks, which correspond to ancient mantles of overthrust. This tectonic and lithomorphic-structural relationship gives a seal of authenticity to the modeling of its relief.

In the center of the Trinidad Mountains, are located the low mountains from 800 to 1 000 m, formed on schist (Topes de Collantes) and those from 900 to 1 200 m, made on carified limestone (Pico San Juan). Surrounding these bigger units, there are small mountains from 500 to 800 m, elaborated on shales, gneisses, limestones and amphibolites.

Both these mountains of Trinidad and those of the block-dome of Sancti Spíritus, constitute a geomorphic unit with precise limits, lower degree of study in comparison with other Cuban mountains, a well-defined altitudinal zonality, a group of geomorphic phenomena grouped in a relatively small area, is currently in process of assimilation and has a good level of knowledge about the historical changes in the natural and socioeconomic environments. Moreover, the extreme meteorological phenomena that have occurred in recent years show their sensitive degree of fragility.

These mountains have an altitudinal zone that is expressed through different landscapes. In the mountainous nucleus, above 800 m of altitude, there are the very humid temperate forests, with annual averages of 2 000 mm of rain and 15º-17ºC of annual average temperature of the air, a stable atmospheric pressure and an annual average relative humidity of 86%. In this area the soils are typical red ferrallitic and brown-red fersialite, which support the rainforest. Towards the edges of the massif are located the temporarily humid forests, with annual average rainfall of 1 900 mm between 700 and 800 m and 1800 mm at 300 m altitude, with brown soils with carbonates, humic, little evolved and skeletal on the southern slope. The forests of this area are the submontane evergreen and the deciduous.

An important peculiarity of these mountains is the presence of a karstic massif located in its central part and in other sectors distributed by its periphery, which gives it a diversity of environments and geomorphological processes.

In these mountains the degree of environmental alteration and, in particular, of the relief due to extreme weather events or during the course of the rainy seasons, has antecedents in the inadequate management of the hydrographic basins, in the ignorance of the structure and operation of the karst reliefs, in its interaction with erosives and in the lack of an effective scheme of protection and conservation of the sub-basins in their diversity, as well as in the problems of design of the works of factory (of drainage and protection) of the highways and roads and, especially, in the absence of essential engineering actions to undertake in the fluvial valleys, fluvio-karst and other depressions of the karst. As a result of the above-mentioned aspects, the negative environmental changes are manifested in a remarkable way in the increase of the intensity of the erosive - accumulative and gravitational processes, in the phenomena of periodic flooding of the fluvial valleys
- karst and other karst depressions and in the numerous alterations occurred in the productive territorial complexes.

Based on estimates and stationary measurements, losses of soil and mineral bark through erosion have doubled and tripled in reforested territories without original forest species and may be higher in deforested areas for agricultural use, with slopes greater than [50]. On the main slopes of the mesorelief, dozens of new or reactivated gravitational forms are observed, with maximum densities of 3 to 10 forms/km2 and, in smaller numbers, on slopes of the mesorelief (with 1 - 2 forms/km2). On the slopes of unprotected excavation roads, new and reactivated or mobilized forms of the order of 2 - 3 forms/km2 were produced as a result of Hurricane "Lili" for certain sections of the main roads (Trinidad - Topes, Topes - Manicaragua, Topes - La Sierrita, among other important ones).

The increase of erosive and gravitational processes in deforested areas with steep slopes has led to an increase in the accumulation of sediments in the subway drainage points of closed valleys, producing local floods of the order of 2 - 17 days and 6 - 25 m high that cover significant socioeconomic areas; where great losses and affectations are produced.

The negative influence of these anthropo-natural geodynamic processes occurring in the mountains is also evident in other peripheral geosystems, where transformations occur in the estuaries and mouths of southern rivers, as well as in the valleys and terrace systems of the North and Northwest, both of which are the result of flooding caused by dangerous flood waves.

The heights of Sancti Spíritus, to the East, are constituted by three units of the relief: the small mountain chains of the Banao Mountain range, from 550 to 800 m, very carified; the premountains on limestone mantles of the Southwest of the region La Ceiba - Pitajones; and the small mountains from 500 to 800 m, on very metamorphosed and folded schists of the high basin of the Caracusey river. The maximum height of this massif is Loma de Banao, with 842 m.

The depression of the Agabama River separates the Trinidad and Sancti Spíritus mountain ranges. It is a sunken tectonic engraving and divided into two units, one of them to the South, formed by a series of erosive and flat accumulative fluvial terraces, and the other to the North, constituted by heights and premountains, precisely involved with the neotectonic ascents of the mechanisms associated to the disjunctive system of Cuban direction. At the same time, the southern end of this depression is blocked, to the South, by chains of low heights, elaborated on monoclinic carbonate sequences.

The essential features of the morphostructure and morpho-sculpture of the mountains of Guamuhaya are directly dependent on their lithological differentiation and the pattern of their palaeo and neo-structural construction.

In this way, the pattern of exogenous modeling on metamorphic schists, reflects embedded hydrographic networks, determining mountain and premountainous chains with sharp crests and steep slopes, formed by very intense erosive processes, often associated with gravitational phenomena. In contrast with the above, on the carbonate rocks there are intensely karst surfaces, with a relief of domes and flat-bottomed depressions.

In the lower part of the southern macro-vertient of the mountain massif, especially towards the Trinidad area, there is a clear spectrum of abrasive, karst terraces that slope gently towards the sea.

In the morphostructures elaborated on methylene substrate, the modelling action of the intense erosion produces morphological modifications in the micro-relief, emerging furrows and gullies. Researches on morphometric dynamics of current erosive forms, carried out by the Institute of Tropical Geography in a station by means of instrumental methods, in Topes de Collantes, between 1984 and 1985, reported an annual growth of 13 cm/year in the width of gullies and 26 cm/year in their headwaters.

The secular and sudden changes caused in the relief by the gravitational and deluge - colluvial processes are clearly expressed in the remarkable alterations of some morphoelements (dividing walls, slopes, erosive levels hanging in the valleys, low erosive and accumulative terraces, etc.) and the morphology of the micro-relief. In localities underlain by carbonate and metacarbonate sequences, they are expressed by circles and cones of landslides and rock falls; while in the sectors modeled on rocks of the terrain or metaterrigenic complexes, landslides and partial or complete packages of weathering barks and soils stand out.

The studies carried out by the Department of Mountains of the Institute of Tropical Geography of the Ministry of Science, Technology and Environment, during more than a decade in the mountainous territories of the country, reveal that the greater frequency in the occurrence of these phenomena of exogenous geomorphological origin, appear in those vertical or subvertical surfaces devoid of the original forest cover and in the slopes of the roads. On the other hand, on the surface of the slopes dedicated to the agricultural and cattle activity without protection practices, the greater concentrations of these destructive forms of the relief are concentrated.

The development of the karst in the Guamuhaya massif has been determined by a number of factors, among which are The complex geological structure resulting from a long process of overthrusting in different directions.

The role of regional metamorphism in the alteration of the original mineralogical properties of the rocks, whose particular characteristics gave rise to the formation of three rings well differentiated by their degree of lithological transformation.

The long period in which the massif has been emerging and the general ascent rates of the region during the process of formation of the relief, responsible for the characteristics of the dissection, the existence of different flattening surfaces and the progressive sinking of the local and regional base levels.

Last but not least, the climatic variations and the predominance of high rainfall throughout the period of formation of the relief probably since the beginning of the Pliocene.

As a result, there are nuclei of ample development of karst forms and, surrounding these, rings where the karst processes are much more discreet, coinciding with the most metamorphosed areas.

Rapidly large volumes of water in the floods, and the emissivities, less inclined in the discharge areas, corresponding with the points where the local base levels cut the surface, constituting the births of many important surface currents of these mountains as is the case of the Hondo River, which is born from one of the most powerful karst sources. In correspondence with the described characteristics, the caves in this mountainous massif have an unequal spatial distribution and do not occupy the totality of its extension. Nevertheless, this typical karst phenomenon is very abundant in Guamuhaya and presents singularities that highlight it from the rest of the country.

First of all, most of the caves in this region are of complex origin, where the processes of erosion and fluvial corrosion have played a preponderant role. The characteristics of the paleogeographic evolution, geological structure and hydrogeological conditions have predetermined the primary morphology of the caves and their water functioning.

Thus, there is a predominance of inclined, staggered, mostly absorbent cavities in the recharging areas where small seasonal currents are submerged but capable of conducting the country's Vauclusian type, located near the town of El Naranjo, municipality of Cumanayagua, province of Cienfuegos. Much less frequent are the transfluent caves (with sinkhole and wheezing) although some stand out like the cave of Los Leones in the ecological reserve Alturas de Banao, Sancti Spíritus, or the cave of Los Tornos in the area of El Naranjo, municipality of

Cumanayagua, Cienfuegos.

The caves of Guamuhaya represent different evolutionary phases according to their geological age, and two large groups can be distinguished in them.

The senile caves, practically inactive from the hydrological point of view, are located in the highest areas, generally near the top of elevations, where they were left hanging due to the effects of dissection and the progressive lowering of the local base levels.

These caves are characterized by the wide development of the classic processes, which have given them gigantic dimensions, as it is the case of the cave of Jose Salas, located in the skirts of Majuana's tits at 900 m of altitude, to the West - Southwest of the town of Avocado, municipality of Cumanayagua, Cienfuegos, to which one enters for an enormous collapse in his vault and that gives access to a room of almost 300 meters of width; or the cave of Martin or Martin-Infierno as it is also known, of 550 m of linear development and 190 m of unevenness, constituted by a group of rooms of enormous dimensions, the most important of which, the furnia of the Hell, has 200 m of length and more than 100 m of width and lodges in its breast the greater stalagmite of America and one of the greater of the world with
47 m high approximately, although some speleologists have reported a height of 60 m. Martin's cave is a National Monument and is proposed as an outstanding natural element of national significance in the national system of protected areas.

Another singular cave of this group is the cave of La Campana, located near the top of the hill of the same name, 2 km to the east of Pico San Juan, which is distinguished for being the karst cave at the highest altitude in Cuba, since it is a little more than 1 000 m of altitude.

The second group refers to active caves, which, as mentioned, can be either absorbent, constituting sinks, or emitting. Absorbent caves generally present great unevenness
such as the case of the cave of Caja de Agua, in the heights of Banao, Sancti Spíritus, with 350 m of unevenness and 5½ km of linear

development, which turns it into the longest cave of Guamuhaya; or the Cuba - Hungary chasm or Magyar chasm, subway stream located in the area of Grones, very close to the cave of José Salas, which with its 475 m of depth is the cave of greater unevenness in Cuba and perhaps the most complex and dangerous for its exploration.

The emissaries are generally caves with less unevenness and many times they are hydrologically connected to absorbent caves, such as the cave of the Resolladero de Yaguanabo, in the cliffs of the eastern edge of the valley of the same name in Cumanayagua, Cienfuegos, which is presumably connected to the cave of the Sumidero del Saúco. In other cases this connection is difficult to assure without evidence of tracers, as in the case of the cave of El Calvo, in El Nicho, Cumanayagua, where the stream of El Negro is born, one of the main sources of the Hanabanilla River.

In all these cases the role of vadose waters is important, responsible for the decoration of these cavities and sometimes for the formation of the cave, as in the case of Rosendo, located near Avocado, a cavity formed by reverse erosion and covered by one of the most beautiful sets of secondary formations in this mountainous massif.

The Guamuhaya Group is a mountainous region that emerged during the last orogeny, located south of the central portion of Cuba. It limits to the north with the heights of Santa Clara, to the east and west with the valleys of Zaza and Arimao rivers respectively, and to the south with the Caribbean Sea. It occupies a surface of 1948 Km2 (approximately 11 % of the mountainous area of Cuba) and has average heights between 700 and 900 m asl, with some heights that exceed 1 000 m, among which San Juan Peak (1 140 m asl) stands out, as the highest point of the massif.

This great unit of the relief is subdivided into the Trinity and Sancti Spíritus Mountains, separated from each other by the depression of the Agabama River. The massif is generally constituted by a nucleus of schistose rocks intensely metamorphosed, very old, folded and altered. On these schists rest fragments of a mantle of recrystallized carbonate rocks, equally metamorphosed and highly carified.
The northern flank of the system is surrounded, first by a band of

amphibolites and gneisses and then by intrusive granite bodies. Further north, the tuffs of the volcanogenic-sedimentary complex extend. To the South the system is in contact with slightly or not very altered carbonate rocks, represented by careous limestones, sandstones, marls, agglomerates and unconsolidated sediments (gravels, sands, clays, silts, etc.)

The dome-block configuration of the Guamuhaya mountain system has been developed, in the neotectonic stage, on the folded base of the Jurassic methylene and metacarbonate complexes. The neotectonic ascent of the system is directly reflected in the relief through the approximately concentric disposition of the leveling surfaces and the altitudinal floors, the design of the fluvial network, the age of the relief, the disposition of the tectonic scarps that limit it and other indicators. Its complicated ancient structure is passively reflected in the relief as carbonated monoclinic tectonic caps and armored mountains that correspond to some overthrusting mantles.

In the center of the Trinidad Mountains, are located the low mountains from 800 to 1000 m asl, formed by schists (Topes de Collantes) and those from 900 to 1200 m, made on carified limestone (Pico San Juan). Surrounding these bigger units, there are small mountains from 500 to 800 m, elaborated on schists, gneisses, carified limestones or amphibolites.

Fig. 3. Vista 3D del macizo Guamuhaya Cienfueguero.
Modelación autores.

The Mountains of Sancti Spíritus are constituted by three units of the relief: the chains of small mountains of the Banao Mountain range, from 550 to 800 m, very carified; the premountains on limestone blankets of the southwest, of the region La Ceiba - Pitajones and the small mountains from 500 to 800 m, on very metamorphosed and folded schists of the high basin of the Caracusey river. The maximum height of this massif is Loma de Banao, with 842 m snm.

The depression of the Agabama River separates the Trinidad Mountains from the Sancti Spíritus Mountains. It is a sunken tectonic block and divided in two units, one of them to the south, formed by a series of erosive and accumulative flat terraces and the other to the north, constituted by heights and premountains. The boundary between these two units is in the vicinity of the town of Meyer. The southern end (within the study area) of the Agabama River depression, is closed by a chain of low carbonate rock heights.

The relief of Guamuhaya responds directly to the lithology and structural characteristics of the mountain system. On the metamorphic schists, the drainage network is deeply embedded, forming mountain and premountain chains of sharp crests and steep slopes, with very intense erosive and denudative processes, often associated with gravitational phenomena; here the valleys are narrow and form small alluvial plains, flat bottomed and generally dissected.

On the carbonated rocks, intensely karst surfaces are formed, with a relief of domes and depressions in the background. In the south of the mountain system, especially towards the Trinidad area, a series of abrasive, carious terraces, gently sloping towards the sea, and dissected by the drainage network, can be observed.

Guamuhaya is a mountainous system of average height, very dissected, where it predominates a relief considered between complex to very complex, of average slope superior to 18° and values of dismemberment of the drainage network greater than 2 000 m/km^2 for the horizontal dissection and of about 300 m for the vertical dissection.

Such a high degree of complexity of the relief and the high values of the morphometric characteristics of this region, restrict the vocation of the territory fundamentally to the forest use and to the coffee economy; in many cases with the application of anti-erosion measures, which protect the territories, especially those of more pronounced slopes, of the loss of horizons of the ground.

However, it is good to clarify that there are areas such as river valley bottoms, the lower part of intramontane depressions and fragments of flat areas, which accept an agricultural use with a certain degree of diversification, but which must be used under adequate control.

The most frequent affectations to the relief are manifested mainly due to exogenous degrading processes such as denudation, erosion, graviclastic or gravitational processes, etc.; which are accelerated or propitiated in some areas by the existence of human impacts such as deforestation, engineering works in general, etc. At present, it can be stated that more than 50% of the area is affected by different degrees of erosion and also the erosion processes associated with the stoniness of soils, are easily developed in the territory by its morphometric characteristics.

These processes, favored by geomorphological peculiarities, are not only linked to the physical loss of soils, but also to other more indirect degrading effects such as the transfer of contaminating substances to lower regions, where the surface flow of chemical substances, derived for example from coffee cultivation, increases in areas with significant slopes, which in turn has had repercussions, for example, in a notable reduction of typical aquatic species in the permanent currents. On the other hand, the increase of erosion affects in a notorious way the balance of the water that is currently supplied to such important sources as the Hanabanilla - Jibacoa Dam.

In general, although the mountainous system of Guamuhaya has a relief of great energy, which restricts its use only to a small number of economic activities, it has a great amount of small areas that, used rationally, can compensate in part for this limitation. The need to harmonize the complex interactions between man and nature, is one of the aspects that support the studies on the environmental characteristics and its relationship with the relief, not only in those places of greater economic assimilation, but even in the less assimilated, as in the less accessible territories, with certain natural or seminatural status, to which they are inherent certain fragility and ecological instability.

The transformation processes that generally take place in this territory are fundamentally associated with the increase in soil and mineral bark erosion. The degree of alteration of the relief and its multiple effects, have antecedents in the inadequate management of the hydrographic basins, in the ignorance of the structure and operation of the relief and in the nonexistence of an effective scheme of protection of sub-basins; as well as in the problems of design of the engineering works of drainage and protection, of highways and roads.

There are certain conditions that favor these processes of transformation or alteration of the relief, such as: deforested or reforested surfaces without species from the original forest, slopes greater than 5°, erosive and erosive-denutive relief, and average annual precipitation greater than 1000 mm.

Accelerated erosive-cumulative processes produce morphometric modifications and erosive forms of furrow and gully type arise. The data on the dynamics of recent erosive forms in the period between October 1984 and the same month in 1985, when the Institute of Geography installed the Topes de Collantes station to measure contemporary erosion processes, reported an increase in the width of the gullies of 13 cm/year and 26 cm/year in their headwaters.

The changes in the relief due to the occurrence of colluvial and deluvial gravitational processes are expressed respectively in the form of landslides and rock, bark and soil slides in the predominantly carbonate rocks, as well as in surface slides of bark, soil and in smaller volume of rocks in terrain materials.

Gravitational processes generate locally significant morphological modifications mainly in the slopes of the basins devoid of the original forest and in the slopes without protection, mainly, in the cuts practiced with road aims. The most notable transformations, where the greatest density of forms is present, coincide with deforested areas where agricultural activities are carried out without protection measures or where agricultural or livestock activities are carried out on impermissible slopes, as well as in areas with deficient forest treatment.

A great number of transformations of the relief, have their basis, among other factors, in the differentiation of this natural component but we must bear in mind other factors such as geomorphological conditions:

The absolute and relative heights of the units of the relief; the spatial distribution and disposition of the basins and intramontane depressions, of their systems of valleys, abrasions, canyons and depressions of subway drainage; the types of processes of morphogenesis; as well as the general morphology and morphometry of the relief; are aspects that play an important role in the quantity, intensity and distribution of rainfall, in the redistribution of winds and their shock surfaces; also exerting their influence on the quantity, intensity, speed and distribution of surface and subway water runoff.

In this mountainous massif, among the many combined factors causing damage, there are effects that reflect very well the influence of the characteristics of the relief as a factor that determines the relations of tension between the natural elements, as well as the conditions for socio-economic development.

Weather

The climate is among the most important factors of physical-geographical differentiation, it has a zonal distribution, which affects the disposition of the remaining natural components, which together constitute the natural conditions of a territory, as well as its potentialities and limitations for the development of one or another type of economic activity.

It presents a climate with temperatures in which the annual average values go from about 20ºC to 26ºC and more, the average relative humidity is high, with averages above 85%, where the daily maximums, generally above 92%, occur at sunrise, while the minimums descend, at midday, to 60 - 70%.

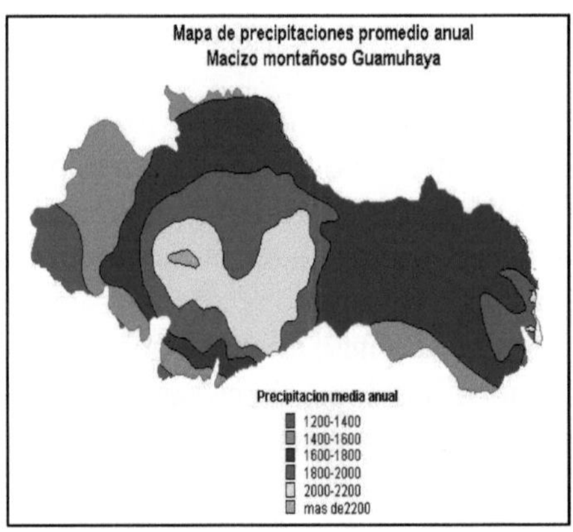

In the area, the winds are predominantly from the east, with speeds between 6 and 10 km / h. The element that varies most in the mountain climate is precipitation, where the annual average exceeds 1000 mm (Saura and Osés, 2003, in Zamora, 2016).

This mountainous massif is located in the northwestern Caribbean sub-region, characterized by variable and calm winds with continental seasonal influence (Alisov dynamic analysis). In Köppen's modified classification it corresponds to the humid tropical climate with rains all the year and warm temperate climate with rains all the year (Barranco, 1989).

The general climatic regionalization of Cuba identifies in the massif the Subtypes 3 and 2 of the mountainous climate with high and stable humidification, low evaporation and fresh temperatures (Díaz, 1989).

Subtype 3:

It extends in the Heights of Sancti Spíritus and in territories with heights lower than 600 m approximately from the Heights of Trinity. This subtype is characterized by:

Table 1: Subtype 3 behavior.

Variable climatic	Indicator	Range
Precipitation ?	Annual average (mm)	1600-1900
	Coefficient of annual variation	0,22-,0,22
	Rainy period (May-October, in % of annual)	75-80
	Average number of days with rain $\chi 1\mu\mu$	90-120
Evaporation	Annual average (mm)	1800-1900
Temperature	Annual average (^{0}C)	23-24
	January (winter, at ^{0}C)	20-21
	July (summer, at ^{0}C)	25-27
Wind ?	Prevailing wind speed (m I sec)	3,3-3,9

Source: (Cutié, et.al, 1999)
*These climatic variables suffer modifications with the height. Precipitation and prevailing wind speed tend to increase; and temperatures tend to decrease as well as evaporation.

Subtype 2:

It is located in the territories with heights higher than 600 m approximately from the Trinidad Heights

Table 2: Subtype 2 behavior.

Variable climatic	Indicator	Range
Precipitation ?	Annual average (mm)	1800-2200
	Coefficient of annual variation	$< 0,2$
	Rainy period (May-October, in % of annual)	60-80
	Average number of days with rain $\chi 1\mu\mu$	100-140
Evaporation	Annual average (mm)	1400-1800
Temperature	Annual average (^{0}C)	16-20
	January (winter, at ^{0}C)	$< 16-20$
	July (summer, at ^{0}C)	$< 20-24$
Wind ?	Prevailing wind speed (m I sec)	3,6-4,2

Source: (Cutié, et.al, 1999)
*These climatic variables suffer modifications with the height. Precipitation and prevailing wind speed tend to increase; and temperatures tend to decrease as well as evaporation.

In general, the maximum wind speeds in this mountain range are associated with frontal systems, extratropical low pressure centers, local storms, cyclonic disturbances and hurricanes; the latter, although less frequent (57 hurricanes of different intensity have affected the territory from 1785 to 1984, 37 of them have occurred between the months of June and November, a period of time known as the Cyclonic Season) are responsible for the maximum values, which are estimated for one. Another characteristic of the wind in this territory is the development of the local valley and mountain regime, and the circulation of strong gravitational winds (Boytel, 1989). Through this mountainous group passes the border that separates the regions of greater and lesser frequency of hurricanes of moderate and great intensity.

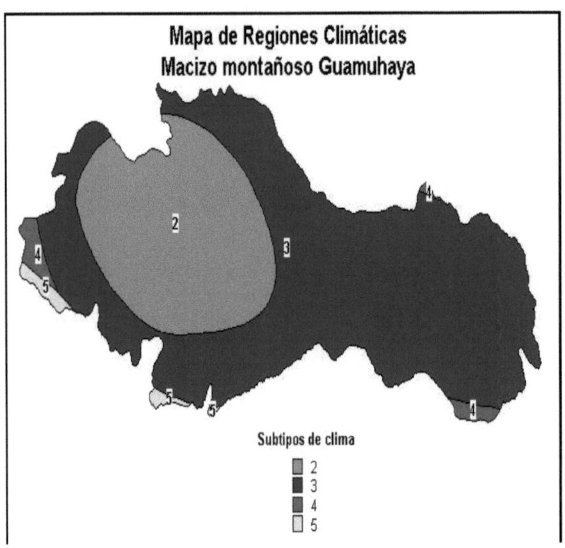

Hydrography

The Agabama River separates the Trinidad and Sancti Spíritus mountain ranges, forming a narrow valley between the two that widens to a 10 km alluvial plain.

The fluvial courses that are born in both mountain ranges tributary their waters to the rivers Agabama, Arimao and Zaza with the exception of those that are developed in their southern slopes.

The Agabama is a river of the southern slope. It is born in the hills of Santa Clara (Sierra de Agabama), it runs 118 kilometers between the mountains of Guamuhaya to the west and the Sierra de Sancti Spíritus to the east. It passes, among others, through the municipality of Fomento and flows into the Caribbean Sea forming a delta; in this final stretch it is called Manatí. It is one of the main rivers in the province of Sancti Spíritus and one of the valleys of the so-called Valle de los Ingenios, a World Heritage Site declared by UNESCO. During the decade of the eighties, last century, were made strongly polluting discharges into this river by the plants Ramon Ponciano, FNTA and Pulp Paper Cuba

The tributaries that drain into the Agabama River are born in the mountains of Trinidad and Sancti Spíritus. In the first one they are located in the oriental part of the north flank and in all its east slope, while in the second one they are developed in the north and west slopes. The main tributaries are the Jibacoa, Caburní and Seibabo rivers.

The Arimao River belongs to the southern slope. It is born in the last foothills of the Escambray mountain range. It irrigates in the beginning the fertile plains of Manicaragua, municipality of Santa Clara and Cienfuegos, and flows to the east of the mouth of Jagua. It has a winding course and measures a total of 80 to 85 kilometers. It favors an important economic development in agriculture and cattle raising, since it supplies water for crop irrigation and for drinking water for the cattle. It has different sandboxes along the pass; you can also fish in the wide pots. Its water is crystalline.

The tributaries of the Arimao River that originate in the heights of Trinidad flow along the western sector of the northern slope and along the entire western slope. Among the important tributaries due to the size of their basins are the Hanabanilla and Mataguá rivers.

The Hanabanilla River (in Cuban aboriginal language: small golden basket) is a tributary of the Arimao River. It is believed that the river receives water from underground aquifers. It is known to form, along with the Guanayara rivers,
Black and Jibacoa the Habanilla reservoir. Until the reservoir, the river runs only 7.3 kilometers of its total of 43.5. Already in 1868 he had

planned to use its waters to supply Cienfuegos, a fact that he legalized in 1908 and materialized in 1911. It has a total of 27 tributaries, which reach distances of between 1 and 4 km. The most important one is the Navarro Stream. The waterfalls known as Salto or Saltos del Hanabanilla are the highest in Cuba, located at 364 m. To the south extends the protected area of Managed Resources of the Hanabanilla sub-basin, where several endemic species such as tocororos, cotorras, carpenters and flowering ducks live.

The Zaza River is 145 kilometers long. It is born in the Guaracabulla area, which belongs to the Placetas municipality in the Villa Clara province, and runs southward to flow into the Zaza dam, which is the largest artificial freshwater lake in Cuba. The riverbed leads to the vicinity of the Tunas de Zaza town, where its dragging has created a large delta.

Zaza comprises 13% of the territory of the province of Villa Clara and 87% of Sancti Spíritus. Its total area is 2,413 km² (2.2% of the national territory). It is estimated that 93% of the soils present different degrees of affectation, 4.5% is very strongly eroded and 16% has been classified as strongly eroded. Deforestation is extensive, only 2.4% of the area is covered by forest. It collects waste from various contaminating sources along its path, as most of it lacks treatment systems. The most problematic belong to the Sugar Industry and the Ministry of Agriculture, and the extension of this river basin measures 2,394 km², exceeded only by the volume and size of Cauto itself.

The rivers Tuinicú, Yayabo and Cayajana are born in the oriental slope of the mountain range of Sancti Spíritus and are tributaries of the Zaza. In the Sierra de Trinidad are located the rivers Gavilanes, San Juan, Yaguanabo, Hondo, Cabagan, Guanayara and Cañas; and in the Heights of Sancti Spirits are located the rivers San Pedro, Higuanojo (Hondo), Los Charcos, Tayabacoa and Banao.

The fluvial courses that are born in both mountain ranges tributary their waters to the rivers Agabama, Arimao and Zaza with the exception of those that are developed in their southern slopes.
The tributaries that drain into the Agabama River are born in the mountains of Trinidad and Sancti Spíritus. In the first one they are located

in the oriental part of the north flank and in all its east slope, while in the second one they are developed in the north and west slopes. The main tributaries are the Jibacoa, Caburní and Seibabo rivers, the Trinidad and Caracusey mountains and Unimazo de las Alturas de Sancti Spíritus. The Agabama River is the geographic feature that separates the Trinidad and Sancti Spíritus mountains, forming a narrow valley between them that widens to 10 km of alluvial plane.

The tributaries of the Arimao River that originate in the heights of Trinidad flow along the western sector of the northern slope and along the entire western slope. Among the important tributaries due to the size of their basins are the Hanabanilla and Matagua rivers.

The rivers Tuinicú, Yayabo and Cayajana are born in the oriental slope of the mountain range of Sancti Spíritus and are tributaries of the Zaza. In the Sierra de Trinidad are located the rivers Gavilanes, San Juan, Yaguanabo, Hondo, Cabagan, Guanayara and Cañas; and in the Heights of Sancti Spirits are located the rivers San Pedro, Higuanojo (Hondo), Los Charcos, Tayabacoa and Banao.

In general, the upper third of rivers have an average annual runoff between 1000 and 1200 mm in the Trinidad Mountains and between 600 and 800 mm in the Sancti Spíritus Mountains, which decrease with altitude, reaching minimum values in the lower thirds of the sub-basins tributary to the main rivers (400-600 mm). Given the presence of the karst, many of these rivers have a subway component in their fluvial runoff, which represents 33.3% of the same in the highest parts of the Guamuhaya massif, weakening its contribution with the decrease in altitude (16.6%) due to the decrease in springs and the predominance of the superficial component of fluvial runoff (Rodríguez, 1989).

The density of the drainage network has a differential distribution in the Massif. On the southern slope of the Trinidad mountains, the basins with the highest density (> 2.5 km/km2) are located towards the west, decreasing to 1.50. In the rest of the Massif, density ranges from 1.00 to 1.50, with the exception of the highest parts of the eastern slope and the southern basins of the Sancti Spíritus Mountains, where values vary between 1.50 and 2.00, as well as in the flood plain and first terrace of the lower third of the Agabama River, which fluctuate between 0.25 and 0.50 (Batista, 1989).

Despite the predominance of forest cover and the presence of karst, the solid modulus of runoff is among the highest in the country and is located in the highest territories of the Trinidad mountains (300-400 t/km2 annual average) and Sancti Spíritus (200-300) decreasing with altitude up to 50 t/km2 annual average.

Some river valleys in the Trinidad Mountains present karst characteristics, their waters drain towards sinks, as is the case of the Boquerones and Jibacoa sub-basins, which form an integrated surface-subterranean (fluvio-karst) drainage system, whose runoff towards the lower course of the Jibacoa River depends on, of the evacuation capacity of the drainage system of the polja (Las Trancas cave) and the magnitude of the contribution of sediments and vegetable matter coming from the extra-area territory (influx), which can obstruct or seal this subway drainage focus, limiting the discharge capacity of the flood waters during the occurrence of intense rains and causing fluvial floods, especially under the conditions of strong slopes of the channels and the morphometric characteristics of these basins, as it happened during the passage of Hurricane Lili on 17 and 18 October 1996, where the erosive - accumulative processes produced morphometric modifications in the headwaters, edges and channels of the erosive forms of lower orders and in the systems of low terraces, Flood planes and riverbeds of the valleys of higher orders, carrying large volumes of sediments and materials, which favored the flood phenomena by the sedimentary obstruction of the conduits of the subway drainage in the blind valleys and poljas (Cuban Center, Jibacoa valley and other depressions). This flooding lasted 17 days.

To the west and center of the Guamuhaya massif there are two manifestations of mineral waters: San Juan and Yaguanabo. The natural emergence expense of these waters reaches 0.5 l/sec and they belong to the physical and fissure-karst type, with pressure.

The San Juan and Yaguanabo manifestations are geologically found within the Trinity-Zaza structural-facial zone, represented by the Naranjo, Narcizo, Mayari, Yaguanabo and Arimao formations, a member of the Mokas.The aquifer complex of the Yaguanabo formation occupies the southeastern part of the area and is formed by epiditotic albino shales and affebotized calcareous shales; the waters of this complex are of fissures and without pressure. Its mineralization and radon content are obtained by a process of eluviation associated with the metronza layer.

The San Juan aquifer complex is made up of limestone and marble with interspersed metamorphosed magmatic rocks and its waters are fissure - karst and phreatic.

There are sulfur springs located in the San Juan River basin, with a flow rate (Q) of 0.1 to 2 l/sec. Among the diseases treated with these waters are: respiratory, skin, osteomuscular and nervous system diseases.

The mountainous group of Guamuhaya is located in the hydrological subregion Sierra del Escambray - Alturas del Norte de las Villas belonging to the Central Region.

The geographical location of this Group, to the south of the central water part of the Island of Cuba, conditions that the fluvial waters drain towards the southern marine platform.

The drainage network is radially distributed, which is determined by the mountainous, anti-form dome morphology, where many of the river courses run along the fault lines. Another feature is the presence of waterfalls such as the Caburní, evidencing the youth of the relief.

The fluvial courses that are born in both mountain ranges tributary their waters to the rivers Agabama, Arimao and Zaza with the exception of those that are developed in their southern slopes.

The tributaries that drain to the Agabama are born in the mountains of Trinidad and Sancti Spíritus. In the first one they are located in the oriental part of the North flank and in all its East slope, while in the second one they are developed in the North and West slopes. The main tributaries are the Jibacoa, Caburní and Seibabo rivers of the Trinidad and Caracusey group and Unimazo de las Alturas de Sancti Spíritus. The Agabama River constitutes the geographic accident that separates the Trinidad and Sancti Spíritus Mountains, forming between both a narrow valley that widens and reaches 10 km of alluvial plane.

The tributaries of the Arimao River that originate in the Trinidad

Mountains flow through the western sector of the North Slope and throughout the West Slope. Among the tributaries of importance for the size of their basins are the Hanabanilla and Matagua rivers.

The rivers Tuinicú, Yayabo and Cayajana are born in the oriental slope of the Mountains of Sancti Spíritus and are tributaries of the Zaza.

Runoff water from the southern slopes of both mountains usually drains into rivers with smaller basins than the previous ones and discharges directly into the sea. In the Trinidad group are the Gavilanes, San Juan, Yaguanabo, Hondo, Cabagán, Guanayara and Cañas rivers; and in the Sancti Spíritus Heights are the San Pedro, Higuanojo (Hondo), Los Charcos, Tayabacoa and Banao rivers.

In general, rivers have in their upper third an average sheet of annual river runoff between 1000 and 1200 mm in the Trinidad Mountains and between 600 and 800 in the Sancti Spíritus Mountains, which decrease with altitude reaching minimum values in the lower thirds of the sub-basins tributary to the main rivers (400-600 mm) (Karasik, 1989a). Given the presence of the karst, many of these rivers have a subway component in their fluvial runoff, which represents 33.3% of the same in the highest parts of the Guamuhaya Massif, weakening its contribution with the decrease in altitude (16.6%) due to the decrease in springs and the predominance of the superficial component of fluvial runoff (Rodríguez, 1989).

The hydrological regime is characterized by its temporality, given the seasonality of rainfall. However, there are many rivers with permanent drainage due to the karst development in the upper third of the basins and the relatively dense forest cover that predominates in this massif, both of which act as regulating agents of runoff. In addition, the hydrographic network in non-karst areas is made up of a group of rivers and streams that in their majority have a permanent character, standing out among them the Hondo, Guanayara, San Juan and Cabagán.

The density of the drainage network has a differential distribution in the massif. The basins with the highest density (> 2.5 km/km2) are located on the southern slope of the Trinidad Mountains, and decrease to 1.50 in the west. In the rest of the massif the density ranges between 1.00 and 1.50, with the exception of the highest parts of the eastern slope and the south basins of the Sancti Spíritus Mountains, where values vary between 1.50 and 2.00, as well as in the flood plane and first terrace of the lower third of the Agabama River that fluctuate between 0.25 and 0.50 (Batista, 1989).

Soils

The soils of Guamuhaya have a close dependence with the lithology, the relief and the climate. According to the genetic classification of Cuban soils, 1971, in the massif four groups of soils predominate widely, which is where we elaborated our map and were able to carry out the work.

The soils that make up the massif are:
Latosolics: Less evolved Tropical
Browns: Typical and Humidified
Limestones: Browns and Reds
Tropical Gleyes: Typical
Soils: Specifically, Red-YellowTypical Mountainous

Within these, Limestone soils actually predominate, occupying about 1309 km2 and constituting 56.7% of the total area of the massif. They are spatially distributed towards the central portion of the territory, the east of it, in a strip to the west and south. Practically these soils could be used more for the minor agricultural cultivation, as well as for the plantain, citrus and fruit trees in general, for having little depth, as well as the existence in many areas of outcrop of calcareous rocks in the surface.

The Tropical Brown soils occupy an area of approximately 500 km2, constituting 21.6% and are located mainly towards the northern end of the territory, with some scattered areas towards the southern end. .these soils for having good fertility are recommended for the cultivation of cane in the deepest places and the use of minor crops in the rest of the areas.

The mountainous soils occupy an area of 347 km2 and constitute 15.03% of the total area of the complex. They are located in the central portion of the Sierra de Trinidad. . They have been traditionally recommended for reforestation due to the type of topography they possess and the need to protect the surface horizons from erosion, although coffee can also be planted and even mixed with the forest plantations.

The Latosolicos cover about 110 km2 and make up 4.7%, distributed in strips towards the western portion of the Guamuhaya group and south of the Sierra de Sancti Spíritus. Only the less evolved soils are presented, which are not fertile, independently they can be used in some places, minor crops, bananas, citrus and others.

Tropical Gley soils cover 42 km2 and constitute 1.8% of the territory. They are located in the form of a strip to the north of the Sierra de Sancti Spíritus and to the south of the Agabama depression, which appears to be included within the territory. It is of great complexity when indicating its agricultural aptitude, due to the diversity of materials that form them and to the topographical position that they occupy.

Table 3. Extension and agricultural aptitude of the soils

Soil type		Area	Percenta ge	Agricultural aptitude
Latosolics (ii)	Less evolved	110 Km²	4,7 %	High productivity, sugar cane, citrus, coffee, bananas and small fruits.
Tropical bales (iv)	Typical and humidified	500 Km²	21,6 %	Appreciable fertility, grasses and minor crops (corn, yucca, etc). The deeper ones can used in cane.
Limestone (vi)	Bards and reds	1309 Km²	56,7%	Citrus and fruit trees, crops minors (chili, tomato, yucca, corn, etc), bananas.
Tropical Gley (viii)	Typical	42 Km²	1,8%	Not recommended for use agricultural
Mountain ous (sa)	Mountain yellowish red typical	347km2	15,03%	Forestry use and according to its coffee altitude.

Fig. Soil area (km2)
of Soil Extent . (%)

Fig. Percentage

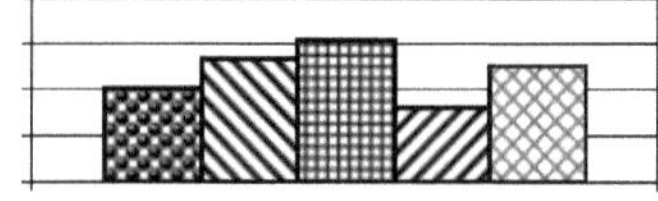

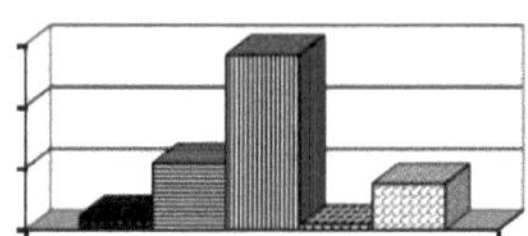

<table>
<tr><td>

■LATOSOLICS   ❑TROPICAL BROWNS
❑LIMESTONES   ❑TROPICAL GLEY
❑MOUNTAINS

</td><td>

■LATOSOLICS   ❑TROPICAL BROWNS
❑LIMESTONES   ❑TROPICAL GLEY
❑MOUNTAINS

</td></tr>
</table>

Biota

One of the distinguishing features of the flora in this territory is the high degree of endemism, whose number of endemic species is around 90, and the variety of plant formations, where secondary vegetation predominates, characterized by forests, shrubs and secondary herbaceous communities, then appear the tropical forests latífolios, represented by the submontane mesophilic, the montane rainforest and low altitude mesophilic.

Institute of Tropical Geography (2001) in Zamora (2016), referring to the forest formations characterizes them as follows: The tropical mountain rainforest develops in a zone of high rainfall at an altitude between 700m and 800m, it is characterized by presenting two arboreal strata, with an abundance of tree ferns and epiphytes in general and a coverage between 90 and 95%. Among the most representative species we have the laurel Ocotea wrigtii, the butterfly Magnolia cubensis ssp, the yagrumite Bocconia frutescens Ricardo (1998), cited by Zamora (2016). In the center of the Sierra de Trinidad, surrounded mainly by the submontane evergreen forest, there is the largest area of this type of forest, as well as small lots, more in the center of the territory, mixed with coffee.

The Mesófilo Submontano Evergreen Forest is located in the Trinidad and Sancti Spíritus mountain ranges. In the first one, associated with the mountain rainforest, typical mesophilic, secondary vegetation and small areas of coffee and forest plantations, and in the second one, with the pastures, secondary vegetation and forest plantations in small proportions. It appears at altitudes between 300 and 800 m. It has two arboreal strata, there are abundant lianas and little development of epiphytes, the vegetation occupies a maximum coverage of 90% and among the species of greater presence appear, the yaya Oxandra lanceolata, the ayua Zanthoxylum martinicense, the jagüey Ficus aurea and the macagua Pseudolmedia spuria Ricardo (1998), cited in Zamora (2016).

On the other hand, this same author is located in the south of the Trinidad mountain range, near the coast, in the extreme southwest, is the Coastal and Sub-Coastal Microphilic Forest, which has two tree strata, with species if empreverdes and deciduous, with some columnar cactus. The vegetation occupies a maximum coverage between 70 and 75%, where you can find the lily Plumeria emarginata, local endemic oak, Tabebuia trinitensis, the seedling Bursera simaruba and bighorn avocado *Dendrocereus nudiflorus.*

With respect to the Typical Mesophile Semideciduous Forest, it is characterized by two arboreal strata and a coverage of up to 70%, it presents abundant shrubs, while herbaceous and lianas are scarce. Among the species that are most abundant are the caguaní Mastichodendro foetidissimum, the dagame Calycophyllum candidissimun and the yaya Oxandra

lanceolata. It is located in the eastern, southern and southwestern part of the Sierra de Trinidad, mainly associated with submontane mesophilic forest, secondary vegetation, coffee, forest plantations, pastures and various crops. In the extreme west of the Sierra de Sancti Spíritus, it appears on a smaller scale associated with pastures, secondary vegetation and various crops (Zamora, 2016).

As far as sugar cane is concerned, it is mainly located in the Agabama basin, in the southern limit of the territory and in a smaller proportion to the north of the Sancti Spíritus mountain range, associated to rice, pastures and various crops, while the various crops are located in the southern head of the Hanabanilla basin and in the northern limit of the massif. Coffee is mainly found in the southeastern part of the Sierra de Trinidad, associated with submontane and typical mesophilic forests, secondary vegetation, pastures and various crops.

The forest plantations stand out mainly for the male pine *Pinus caribaea, Eucalyptus sp,* the blue majagua *Hisbiscus elatus and* the casuarina Casuarina equisetifolia, introduced by man, which may appear mixed or be planted with coffee in its understory. These plantations are distributed throughout the Sierra de Trinidad, mainly in the central and southeastern part, as well as in the Sierra de Sancti Spíritus, located in small lots (Zamora, 2016).

With respect to the fauna of the massif (IGT, 2001 in Zamora, 2016), he considers that it is rich and diverse, spread over the three regions that make up the massif. In the mountainous zone there is a varied animal world in which three endemic mammals of the country stand out: the hutia conga *Capromys pilorides, the* hutia carabalí *Capromys prensilis,* arbolícola and the bat of hot caves *Phyllonycteris poeyi.*

The bird life is very diverse, among the endemic species abound: the fart or cartacuba *Todus multicolor,* the tocororo *Priotelus temnurus,* Cuba's national bird, two species of woodpecker: the scapular *Colaptes auratus* and the jabao *Melanerpes superciliaris,* the first flies in the xerophytic forests of the southern slope, while the second is the most abundant of Cuban woodpeckers. You can also find the *Psittacara euops* cat, two species of siju: the *Glaucidium siju* banana tree and the *Otus lawrenci* cotunto. The endemic subspecies in the arboreal spaces of this locality: the *Amazon* parrot *leucocephala.*

Among the birds, there are other endemic ones such as the *Patagioenas cyanocephala* dove, the longtailed hawk *Accipiter gundlachi* and the bustard *Teretistris fornisi,* among others. They populate the forests of the mountains
13 species and 19 subspecies of endemic and 33 non-endemic birds In addition, there are 7 species of anuran amphibians in the highest areas, standing out in Collantes Topes the chubby little vulture *Eleutherodactilus emiliae* and the Collantes Topes toad *Peltophryne dunni.* Endemic reptiles of the Saurian suborder, such as *Leiocephalus, Chamaeleolis and Anolis,* are abundant. From the suborder ophdea are found, the maja of Santamaria *Chylabotrhus angulifer and* the jubo de magdalena or *prieto Antillophis andreae and* the jubo sabanero *Alsophis cantherigerus.*

MATERIALS AND METHODS

The bibliographic information is considered updated until June 30, 2020. The order in which the taxa are presented is only alphabetical. The examined material is deposited in the Institute of Ecology and Systematics (IES), the National Museum of Natural History (MNHNC), Havana, and the personal collection of Giraldo Alayón García (GAG), San Antonio de los Baños, Havana.

Reflecting the geographic distribution of each species in the Guamuhaya Mountains, the type locality is underlined. Species that constitute new records are identified as (NRE); and new locality records as (NRL).

The records for "Soledad" (now Cienfuegos Botanical Garden) have not been included because its eastern portion sits in the foothills of Guamuhaya and, moreover, many of the spider records assigned to that location may correspond to originally forested areas of these hills. The records corresponding to the city of Trinidad have also been excluded, since it is located at the foot of the Trinidad Heights, near its northern part.

Several new records correspond to "Boca de Carreras" and "Agua Hedionda", two locations located very close to each other, south of Pico San Juan and not far from Buenos Aires (21º 58' N - 80º 09' W), municipality Cumanayagua, province Cienfuegos. Unfortunately, on the date of the surveys (August 1978), no satellite georeferencing equipment was available and in some publications these locations have been described as belonging to the province of Sancti Spíritus (Armas & Juarrero, 1999; Alayón, 2000) or have not been properly located (Alayón, 2003).

RESULTS

ARANEA ORDER
Suborder Mygalomorphae Family Barychelidae
1. *Psalistops sp.* Cienfuegos: Cumanayagua: José Salas Cave (Ramos & Hernández, 2007). Threshold species.
2. *Trichopelma sp.* Cienfuegos: Cumanayagua: Cueva Amistad Cuba-Magyar, Cueva de Rosendo, Cueva de Las Aromas, Cueva del Vencejo (Ramos & Hernández, 2007). Troglodyte species.

Dipluridae Family
3. *Euagrus sp.* Cienfuegos: Cumanayagua: José Salas Cave (Ramos & Hernández, 2007). Troglodyte species.
4. *Ischnothele longicauda* Franganillo, 1930 (NRL). Sancti Spíritus: Trinidad: Arroyo Caburní, Topes de Collantes. Alayón (1994) had mentioned it from Guamuhaya, but without indicating a precise location.

Family Theraphosidae
5. *Cyrtopholis bryantae* Rudloff, 1995. Cienfuegos: Cumanayagua: San Blas. Species erroneously identified by Bryant (1940:261-262) as *Sphaerobothria gibbosus* (Franganillo, 1935). Local endemism.
6. *Phormictopus sp.* Cienfuegos: Cumanayagua: José Salas Cave (Ramos & Hernández, 2007).
7. *Euripelma spinicrus.* Sancti Spíritus: Lomas de Banao.

Suborder Araneomorphae Family Amaurobiidae
8. *Tugana cudina* Alayón, 1992. Sancti Spíritus: Trinidad: Cudina, Topes de Collantes. Local endemism; lives in limestone cliffs (Alayón, 1992, 2000). The typical locality is also known as Finca Codina.

Family Anyphaenidae
9. *Hibana fusca* (Franganillo, 1926). Sancti Spíritus: Trinidad: Topes de Collantes (Alayón, 2000). National endemism.

10. *Wulfila longipes* (Bryant, 1940). Cienfuegos: Cumanayagua: Buenos Aires (Bryant, 1940; as *Anyphaenella longipes*).

Family Araneidae

11. *Argiope argentata* (Fabricius, 1775). Alayón (1994) mentioned it from Guamuhaya, without indicating a precise location.
12. *Argiope trifasciata* (Förskal, 1775). Alayón (1994) mentioned it from Guamuhaya, without indicating a precise location.
13. *Cyclosa caroli* (Hentz, 1850) (NRE). Cienfuegos: Cumanayagua: Boca de Carreras.
14. *Cyclosa turbinata* (Walckenaer, 1842). Alayón (1994) mentioned it from Guamuhaya, without indicating a precise location.
15. *Cyclosa walckenaeri* (O. P.-Cambridge, 1889) (NRE). Sancti Spíritus: Trinidad: Topes de Collantes.
16. *Eriophora ravilla* (C. L. Koch, 1844). Cienfuegos: Cumanayagua: Boca de Carreras (NRL).
17. *Eustala anastera* (Walckenaer, 1842). Alayón (1994) mentioned it from Guamuhaya, without indicating a precise location.
18. *Eustala fuscovittata* (Keyserling, 1864) (NRE). Sancti Spíritus: Trinity: Tops of Collantes.
19. *Metazygia zylloides* (Banks, 1898) (NRE). Cienfuegos: Cumanayagua: Boca de Carreras.
20. *Metepeira triangularis* Franganillo, 1930. Alayón (1994) mentioned it from Guamuhaya, without indicating a precise location. National endemism.
21. *Micrathena forcipata* (Thorell, 1859). Cienfuegos: Cumanayagua: Buenos Aires (Bryant, 1940) Boca de Carreras (NRL).
22. *Micrathena horrida* (Taczanowski, 1873). Alayón (1994) mentioned her from Guamuhaya, without indicating a precise location.
23. *Micrathena militaris* (Fabricius, 1775). Cienfuegos: Cumanayagua: Boca de Carreras (NRL).
24. *Neoscona marcanoi* Levi, 1993. Alayón (1994) mentioned it from Guamuhaya, without indicating a precise location.
25. *Neoscona moreli* (Vinson, 1863). Alayón (1994) mentioned her from Guamuhaya as *N. neotheis* Petrunkevitch, 1911, but without indicating a precise location.
26. *Neoscona nautica* (C. L. Koch, 1875). Alayón (1994) mentioned it from Guamuhaya, without indicating a precise location.

27. *Verrucosa arenata* (Walckenaer, 1842). Sancti Spíritus: Trinidad: Topes de Collantes: La Chispa (Ballesteros) (NRL) and Caburní River (NRL). Cumanayagua: Boca de Carreras (NRL); Agua Hedionda (NRL).

28. *Witica crassicaudus* (Keyserling, 1865). Sancti Spíritus: Trinity: Collantes' Stops: La Chispa (Crossbowmen) (NRL). Cienfuegos: Cumanayagua: Boca de Carreras (NRL). Alayón (1994) had mentioned it from Guamuhaya, but without indicating a precise location.

Family Caponiidae

29. *Nops sp.* Cienfuegos: Cumanayagua: José Salas Cave (Ramos & Hernández, 2007). Troglodyte species.

Family Clubionidae

30. *Elaver carlota* (Bryant, 1940). Cienfuegos: Cumanayagua: Mina Carlota (as Clubiona carlota Bryant, 1940). National endemism.

31. *Elaver crinophora* (Franganillo, 1934). Sancti Spíritus: Trinidad: La Chispa, Topes de Collantes (Alayón, 2000). Villa Clara: Manicaragua: Salto del Hanabanilla (Bryant, 1940; as Clubiona crinofora). National endemism.

32. *Cuban Elaver* Roewer, 1951 Cienfuegos: Cumanayagua: around El Nicho (Alayón, 2000). National endemism.

Family Dysderidae

33. *Trachelas contractus* Platnick & Shadab, 1974. Cienfuegos: Cumanayagua: Buenos Aires (Bryant, 1940; as *Trachelas bicolor* Keyserling, 1887). Only known from the type locality (Alayón, 2000).

Family Ctenidae

34. *Celaetycheus cabriolatus* Franganillo, 1930 Cienfuegos: Cumanayagua: Buenos Aires (Bryant, 1940)

35. *Ctenus anclatus* Franganillo, 1931. Sancti Spíritus: Trinidad: Topes de Collantes (Alayón, 2000). National endemism.

36. *Ctenus brevitarsus* Bryant, 1940. Sancti Spíritus: Trinidad: Topes de Collantes (Alayón, 2000). Cienfuegos: Cumanayagua: Mina Carlota (Bryant, 1940). National endemism.

37. *Ctenus ramosi* Alayón, 2002. Cienfuegos: Cumanayagua: Rosendo

Cave; Los Mapas Cave (Ramos & Hernández, 2007), Cave
Martin Hell (NRL). A troglodytic species, so far only found in these
three caves. The male is unknown.

38. *Ctenus sp.* Cienfuegos: Cumanayagua: Rosendo Cave (Ramos &
Hernández, 2007). Troglodyte species.

39. *Cupiennius cubae* Strand, 1910. Cienfuegos: Cumanayagua: Boca de
Carreras (NRL)

Family Drymusiidae

40. *Drymusa spectata* Alayón, 1981. Sancti Spíritus: Trinidad: La Chispa,
Topes de Collantes. Cienfuegos: Cumanayagua: Sima Santiago,
José Salas Cave, Cueva Amistad Cuba-Magyar, Rosendo Cave,
Cueva de Las Aromas, Cueva el Vencejo (Ramos & Hernández,
2007); Cueva Las Palmas (NRL). Endemic to the Guamuhaya
Mountain Range, with high humidity requirements; it lives between
300 and 700 meters above sea level (Alayón, 2000).

Filistatidae Family

41. *Filistata hibernalis* Hentz, 1842. Cienfuegos: Cienfuegos:
Guamuhaya (NRL). Alayón (1994) had mentioned her from
Guamuhaya, but without indicating a precise location. Synanthropic
species, common in houses.

42. *Filistatoides insignis* (O. P.-Cambridge, 1896). Alayón (1994)
mentioned it from Guamuhaya, without indicating a precise location.

Family Gnaphosidae

43. *Cubanopyllus inconspicuus* (Bryant, 1940). Cienfuegos: Genus and
species endemic to Cuba (Alayón 2000).

Family Hersiliidae

44. *Yabisi habanensis* (Franganillo, 1936) Cienfuegos: Cumanayagua:
Mina Carlota (cited by Bryant, 1940:274-275, as *Tama habanensis*)
was until recently considered a Cuban endemism, but Rheims &
Brescovit (2004) recorded it from the southern US.

Family Lycosidae

45. *Arctosa fusca* (Keyserling, 1876). Cienfuegos: Cumanayagua: Boca de Carreras. This species seems to be well distributed, at least in the western half of the country (Bryant, 1940).

46. *Pardosa sp.* Cienfuegos: Cumanayagua: Cueva Amistad Cuba-Magyar (Ramos & Hernández, 2007). Threshold species.

47. Gertsch *Mayan pirate,* 1940. Alayon (1994) mentioned it from Guamuhaya, without indicating a precise location.

48. *Pirate sedentarius* Montgomery, 1904. Alayón (1994) mentioned it from Guamuhaya, without indicating a precise location.

Mimetidae Family

49. *Mimetus hesperus* Chamberlin, 1922. Villa Clara: Manicaragua: Salto del Hanabanilla (Bryant, 1940).

Family Mysmenidae

50. *Mysmenopsis tibialis* (Bryant, 1940). Sancti Spíritus: Sierra de Banao (Alayón, 2000). National endemism. This species lives mostly as a kleptoparasite in the webs of *Ischnothele longicauda* (Dipluridae) (Alayón, 2000).

Family Nephilidae

51. *Nephila clavipes* Linnaeus, 1767. Sancti Spíritus: Banao (NRL). Trinity: Collantes Stops (NRL). Alayón (1994) had mentioned it from Guamuhaya, but without indicating a precise location.

Nesticidae Family

52. *Gaucelmus augustinus* Keyserling, 1884 (first record for Cuba). Cienfuegos: Cumanayagua: José Salas Cave, Cueva Amistad Cuba-Magyar, Cueva de los Mapas, Cueva de Rosendo, Cueva del Vencejo (Ramos & Hernández, 2007, as Gaucelmus *sp. n.*); *Cueva* Los Gambusinos: a female (MNHNC); Cueva Eulalio: a female (MNHNC); Cueva de Adriano: a female (MNHNC), on the walls. Troglodyte species, widely distributed in North America, Central America and some Antillean islands. A female (MNHN-C), collected by J. M. Ramos between November 11 and 12, 2000, in Cueva de Falcó, Boquerones Subterranean System, Yaguajay municipality, Sancti Spiritus province, has also been examined.

Family Oonopidae

53. *Ischnothyreus peltifer* (Simon, 1891). Alayón (1994) mentioned it from Guamuhaya, without indicating a precise location.

Pholcidae Family

54. *Modisimus elevatus* Bryant, 1940. Cienfuegos: Cumanayagua: Cueva Las Palmas (NRL): a female, collected in the wall of the penumbras zone. Regional endemism, frequent in caves.

55. *Modisimus sp.* Cienfuegos: Cumanayagua: José Salas Cave, Rosendo Cave, Las Aromas Cave, Swift Cave (Ramos & Hernández, 2007); Los Gambusinos Cave: a juvenile (MNHN-C). Threshold species.

Family Pisauridae

56. *Dolomedes Guamuhaya* Alayón, 2003. Cienfuegos: Cumanayagua: "Boca de Carreras, Sierra del Escambray". Alayón (1994) mentioned it from "Guamuhaya" as *Dolomedes fuscus* Franganillo, 1930. Aquatic species, local endemism.

Note: Bocas de Carreras is located south of Pico San Juan, not far from Buenos Aires.

Salticidae Family

57. *Agobardus cubensis* (Franganillo, 1935). Cienfuegos: "Solitude and Mountains of Trinidad" (Bryant, 1940). National endemism.

58. *Agobardus mandibulatus* Bryant, 1940. Cienfuegos: Cumanayagua: Buenos Aires and Mina Carlota; Cienfuegos: Soledad (Bryant, 1940). It seems to be an endemism of the western half of Cuba (Alayón, 2000).

59. *Blessed Wickhami* (Peckham & Peckham, 1894). Cienfuegos: Cumanayagua: "Mountains of Trinidad, Mina Carlota" (Bryant, 1940)

60. *Corythalia emertoni* Bryant, 1940. Cienfuegos: Cumanayagua: Mina Carlota (Bryant, 1940). Endemism of the western half of Cuba. The type locality was originally designated as "Soledad, Mina Carlota", but this is one of the many cases in which Soledad appears only as a reference point, for being at that time the best known place by the North Americans who used to collect in the lands of the sugar mill "Soledad" (currently Pepito Tey) and its surroundings.

61. *Phidippus audax* (Hentz, 1845). Alayón (1994) mentioned it from Guamuhaya, without indicating a precise location.

62. *Sidusa inconspicuous* Bryant, 1940. Cienfuegos: Cumanayagua: Buenos Aires. It is only known of the type locality.
63. *Sidusa turquinensis* Bryant, 1940. Cienfuegos: Cumanayagua: José Salas Cave (Ramos & Hernández, 2007). Threshold species. It is known from several locations in the eastern half of Cuba (Alayón, 2000).
64. *Synemosyna smithi* Peckham & Peckham, 1893 Cienfuegos: Cumanayagua: Mina Carlota; Cienfuegos: Soledad (Bryant, 1940)
65. *Thiodin inerma* Bryant, 1940. Cienfuegos: Cienfuegos: Loneliness. National endemism.

Family Scytodidae
66. *Scytodes soft* Bryant, 1940. Cienfuegos: Cumanayagua: Agua Hedionda (Alayon, 2000:8 quoted from "Aguas Hediondas, Sierra de Trinidad, province of Sancti Spiritus"). Cienfuegos: Solitude (various records) (Bryant, 1940). This species is mostly arboreal and constitutes a national endemism.
67. *Scytodes cubensis* Alayon, 1977. Sancti Spíritus: Trinidad: Entronque de Topes de Collantes (Alayón, 1977). National endemism.
68. *Scytodes longipes* Lucas, 1844. Alayon (1994) mentioned it from Guamuhaya, without indicating a precise location.
69. *Scytodes noeli* Alayon, 1977. Cienfuegos: Cumanayagua: José Salas Cave (Ramos & Hernández, 2007). Adriano Cave: two females (MNHN-C), on the walls of the threshold zone. (Ramos & Hernández, 2007). Troglodyte species, although more abundant in the threshold zone; its known distribution includes the Sierra de los Órganos (Pinar del Río province) and the Guamuhaya Mountains (Alayón, 2000).
70. *Scytodes robertoi* Alayon, 1977. Alayón (1994) mentioned it from Guamuhaya, without indicating a precise location; later, Alayón (2000) quoted it from "Sierra de Trinidad, province of Sancti Spíritus".

Family Selenopidae
71. *Selenops aequalis* Franganillo, 1935. Sancti Spíritus: Trinidad: Topes de Collantes: El Chorrito (NRL). Alayón (2000) mentioned it as widely distributed in the Guamuhaya Mountains, but without indicating a precise location. National endemism.
72. *Selenops aissus* Walckenaer, 1837 (NRE). Sancti Spíritus: Trinidad: road from Trinidad to Topes de Collantes, km 2

73. *Selenops submaculosus* Bryant, 1940. Sancti Spíritus: Trinidad: 1 km
    E of Trinidad (NRL). Cienfuegos: Cienfuegos: Solitude.

Family Sicariidae

74. *Cuban Loxosceles* Gertsch, 1953. Alayón (1994) mentioned it from
    Guamuhaya, without indicating a precise location.

Family Sparasidae

75. *Heteropoda venatoria* (Linnaeus, 1767). Alayón (1994) mentioned it
    from Guamuhaya, without indicating a precise location. Pantropical
    species.

Family Tetragnathidae

76. *Alcimosphenus licinus* Simon, 1895 Alayón (1994) mentioned it from
    Guamuhaya, without indicating a precise location.
77. *Leucauge argyra* (Walckenaer, 1862). Alayón (1994) mentioned it
    from Guamuhaya, without indicating a precise location.
78. *Leucauge regnyi* (Simon, 1897). Sancti Spíritus: Trinity: Caburni
    Creek, Collantes Tops (NRL). Cienfuegos: Cienfuegos: Solitude
    (Bryant, 1940).
79. *Tetragnatha tenuissima* O. P.-Cambridge, 1889. Cienfuegos:
    Cienfuegos: "Soledad, Montañas de Trinidad"; "Soledad, colina de
    Vilches" (Bryant, 1940).

Family Theridiidae

80. *Anelosimus jucundus* (O. P.-Cambridge, 1896). Sancti Spíritus:
    Trinity: Caburní stream, Topes de Collantes (NRL). Alayón (1994)
    had mentioned it from Guamuhaya, but without indicating a precise
    location.
81. *Argyrodes fictilium* (Hentz, 1850). Alayón (1994) mentioned it from
    Guamuhaya, without indicating a precise location.
82. *Latrodectus geometricus* C. L. Koch, 1841. Alayón (1994) mentioned
    it from Guamuhaya, without indicating a precise location.
83. *Theridion evexum* Keyserling, 1884. Cienfuegos: Cienfuegos:
    Soledad; Cumanayagua: Buenos Aires (Bryant, 1940; as *Theridion
    cabriolatum* Franganillo, 1930).
84. *Theridula gonygaster* (Simon, 1873). Cienfuegos: Cienfuegos:
    "Solitude and Mountains of Trinidad" [Bryant, 1940; as Theridula
    opulenta (Walckenaer, 1837)]. Cumanayagua: Bocas de Carreras

(NRL).

85. *Thymoites levii* Gruia, 1973. Sancti Spiritus: Trinidad: Arroyo El Chorrito, Topes de Collantes (Gruia, 1983). Endemism of the western half of Cuba.

Theridiosomatidae Family

86. *Mexican Wendilgarda* Keyserling, 1886 Alayón (1994) mentioned her from Guamuhaya, without indicating a precise location.

Family Uloboridae

87. *Uloborus trilineatus* Keyserling, 1883 Alayón (1994) mentioned it from Guamuhaya as *Uloborus penicillatus* Simon, 1891, but without indicating a precise location.
88. *Zosis geniculatus* (Olivier, 1879). Alayón (1994) mentioned it from Guamuhaya, but without indicating a precise location. Pantropical species, common in all Cuba.

ENDEMISM AND CONSERVATION

Of the 88 spider species registered so far in Guamuhaya, 28 are clearly restricted to this area, to which should be added another 13 that, so far, have only been found in this massif and its foothills, although it is possible that they have a more extensive geographical scope (Table 4, Appendix I). Other 18 species constitute regional endemisms, for what, globally, the third part of the inventoried species in this orographic system represents Cuban endemisms.

Table 4. Taxonomic composition and endemism of the arachnofauna from Guamuhaya Mountains, central Cuba. The endemism has been classified in local (L), regional (R) and national (N). Fam: Families; Gen: Genera; Esp: Species. Endemism.

TAXONS					Endemism		
Orders	Suborders	Families	Genres	Species	L	R	N
Araneae(1)	2	31	63	88	28	13	24

While it is true that at least in the case of the spiders (Araneae) some of the species supposedly exclusive to Guamuhaya may have a wider geographical distribution, it is also undeniable that the diversity of this fauna in the area is much greater. The study of the soil fauna, constituted by several families whose representatives are characterized by their small size, such as the Oonopidae, Tetrablemmidae, Gnaphosidae and Ochyroceratidae, among others, could considerably expand this first inventory. Caves and forest canopies may also be home to species that are still unknown. Opilions and pseudoscorpions have hardly been studied in this massif, so the present inventory does not adequately reflect their specific richness and diversity either.

While those of 28 other species are distributed mainly among Trinidad, Mina Carlota, Buenos Aires and Topes de Collantes. Globally, this massif is the type locality of almost a quarter of all the species of its arachnophauna. To which it must be added the fact that approximately the sixth part of the species that integrate this fauna only has been found in its properties.

Despite the fact that relevant research has not been carried out to objectively evaluate the conservation status of such fauna, at least those species whose distribution is clearly restricted to a single locality could be included in one of the risk or threat categories.

DISCUSSION

The arachnophauna that populates this massif of central Cuba is a combination of relatively old elements (evidenced by the presence of genus and groups of endemic species) and others that have a close evolutionary and biogeographical relationship either with the eastern or western region of the country, or with more distant faunas, such as the Central American one.

Although the present inventory is not adequate to obtain an accurate view of the taxonomic composition and geographic distribution of the spiders that populate the Guamuhaya Mountains, the results represent a valid approximation. Based on the available information, the arachnophobia of this massif is one of the most important in the country.

ACKNOWLEDGEMENTS

We thank the doctors in biological sciences Luis F. de Armas Chaviano from the Institute of Ecology and Systematics of Havana (IES-CITMA) and Giraldo Alayón García from the National Museum of Natural History of Cuba (MNHNC); for the description of new species for science and the determination of new records of infra-generic taxa for the area. To René Barba and Aylín Alegre Barroso (IES), for the information kindly provided about Pseudoscorpions and Opiliones, respectively.

REFERENCES

ALAYÓN GARCÍA, G. 1992. The genus Tugana (Araneae: Amaurobiidae). Poeyana, 416: 1-7.

ALAYÓN GARCÍA, G. 1994. List of Cuban spiders (Arachnida: Araneae). AvaCient, Mexico, 10: 3-30.

ALAYÓN GARCÍA, G. 2000. The endemic spiders of Cuba (Arachnida: Araneae). Iberian Magazine of Arachnology, 2: 1-48.

ALAYÓN GARCÍA, G. 2002. Notes on the family Ctenidae in Cuba with the description of a new species of *Ctenus* from a cave and the female of *C. conaxus* Bryant (Arachnida: Araneae). Iberian Magazine of Arachnology, 6: 135-139.

ARMAS, L. F. 1995. Taxonomic diversity of Cuban arachnids. *Cocuyo*, Havana, 3: 10-11.

ARMAS, L. F. DE & G. ALAYÓN GARCÍA 1984. Synopsis of Cuban cave arachnids (except mites). Poeyana, 276: 1-25.

ARMAS, L. F. DE & R. TERUEL 2005. The Solifuges of Cuba (Arachnida: Solifugae). Bulletin of the Aragonese Entomological Society, 37: 149- 163.

ARMAS, L. F. DE, G. ALAYÓN GARCÍA & J. M. RAMOS HERNÁNDEZ. 2009. Aracnofauna (except Acari) dlas Montañas Guamuhaya, Central Cuba First Approximation. *Bol. Entomol Soc. Arag.* (45): 135- 146.

BARRANCO, G. & L. R. DÍAZ 1989. Climate regionalization and climate types. In: New National Atlas of Cuba. Institute of Geography of the

Cuban Academy of Sciences, Cuban Institute of Geodesy and Cartography and National Geographic Institute (Spain). Spain VI. 1. 2: 1.

BRYANT, E. B. 1940. Cuban Spiders of the Museum of Comparative Zoology. Bulletin of the Museum of Comparative Zoology, 86(7): 249- 532.

GRUIA, M. 1973. On some Theridiidae (Araneae) collected by biospecies expeditions to Cuba. Results of the Cuban-Romanian biospeological expeditions in Cuba. Edit. Academiei Republicii Socialiste România, Bucharest, 1: 305-316.

ITURRALDE-VINENT, M. A. 1988. Geological aspects of Cuban biogeography. Earth and Space Sciences, 5: 85-100.

RAMOS HERNÁNDEZ, J. M. & A. HERNÁNDEZ MUÑOZ 2007. Results biological of the joint Cuban-Swiss speleological expeditions to the town of Aguacate, Guamuhaya Massif, Cuba. II National Symposium of Biospeleology, Sancti Spíritus. CD ROOM Memories of the National Symposiums of Biospeleology, Feijóo Publishing House, Santa Clara, Cuba. ISBN 959-250-292-7.

RHEIMS, C. A. & A. D. BRESCOVIT 2004. Revision and cladistic analysis of the spider family Hersiliidae (Arachnida, Araneae) with emphasis on Neotropical and Nearctic species. Insect Syst. Evol., 35: 189-239.

SAMEK, V. 1973. Phytogeographical regions of Cuba. Academy of Sciences of Cuba, Forestry series, 15: 1-60, 1 map.

APPENDIX I

List of examined material that represents new locality record for a species (NRL) or new species record for Guamuhaya (NRE)

ARANEA ORDER

Suborder

Mygalomorphae

Family Dipluridae

Ischnothele longicauda Franganillo. Sancti Spíritus: Trinidad: A female, Caburní stream, Topes de Collantes, August 1978, G. Alayón (IES-3.1595).

Suborder

Araneomorphae

Family Araneidae

Cyclosa caroli (Hentz, 1850) (NRE). Cienfuegos: Cumanayagua: A Female, Boca de Carreras, August 1978, G. Alayón (IES-3.1557).

Cyclosa walckenaeri (O. P.-Cambridge) (NRE). Sancti Spíritus: Trinidad: A female and a juvenile male, Topes de Collantes, August 1978, G. Alayón (IES-3.1591 y 3.1592).

Eriophora ravilla (C. L. Koch) (NRL). Cienfuegos: Cumanayagua: Three females, Boca de Carreras, August 1978, G. Alayón (IES-3.1584 to 3.1586).

Eustala fuscovittata (Keyserling) (NRE). Sancti Spíritus: Trinity: a female, Topes de Collantes, September 7, 1972, L. F. de Armas (IES-3.1523).

Metazygia zilloides (Banks) (NRE). Cienfuegos: Cumanayagua: A female, Boca de Carreras, August 1978, G. Alayón (IES-3.2245).

Micrathena forcipata (Thorell, 1859) (NRL). Cienfuegos: Cumanayagua: A female, Boca de Carreras, August 1978, G. Alayón, 330 masl (IES-3.1556); a pre-adult female and a male, both in the same logging, same data as the previous one (IES-3.1558 and 3.1559).

Micrathena militaris (Fabricius) (NRL). Cienfuegos: Cienfuegos: A female, La Legua, 13 km S of Soledad, August 1978, L. F. de Armas (IES-3.2335). Cumanayagua: Seven females, Boca de Carreras, August 1978, G. Alayón (IES3.1560 al 3.1566).

Verrucosa arenata (Walckenaer) (NRL) Sancti Spíritus: La Sierpe: A female, Javico, Sierra de Banao, February 4, 1994, Blas Pérez, in coffee plantation (IES). Trinidad: Three Females, La Chispa (Ballesteros), Topes de Collantes, July 22, 1977, G. Alayón, in the forest (IES-3.2351 al 3.2353); Arroyo Caburní: One Female, August 1978, G. Alayón (IES- 3.1590). Three females, journey from Salto de Vega Grande to Salto del Caburní, Topes de Collantes, December 6, 1991, a. Ávila C. (IES). Cienfuegos: Cumanayagua: A female, Agua Hedionda, August 1978, L. F. de Armas, in semideciduous forest (IES-3.2253). Four females and one male, Boca de Carreras, August 1978, G. Alayón, 330 masl (IES-3.1579 al 3.1583).

Witica crassicaudus (Keyserling) (NRL). Sancti Spíritus: Trinidad: Four females, La Chispa (Crossbowmen), Topes de Collantes, July 1977, G. Alayón (IES- 3.2351 al 3.2354). Cienfuegos: Cumanayagua: Four females and three juveniles, Boca de Carreras, August 1978, G. Alayón, near streams, 330 masl (IES-3.1570 al 3.1576).

Family Ctenidae

Cupiennius cubae Strand (NRL). Cienfuegos: Cumanayagua: One female, Boca de Carreras, August 1978, G. Alayón, 330 m, in bush stratum (IES-3.2240); two females with ootheca, same data as above (IES-3.1568 and 3.1569).

Family Deinopidae

Deinopis lamia Mac Leay (NRL). Cienfuegos: Cumanayagua: A Female, Boca de Carreras, August 1978, G. Alayón (IES3.1588).

Family Drymusidae

Drymusa spectata Alayón (NRL). Cienfuegos: Cumanayagua: Two females, Cueva Las Palmas, March 5, 2007, J. M. Ramos, on walls of the threshold zone (MNHN-C).

Filistatidae Family

Filistata hibernalis Hentz (NRL). Cienfuegos: Two females, Soledad, September 1975, I. García (IES-3.1516 y 3.1517).

Family Nephilidae

Nephila clavipes Linnaeus (NRL). Sancti Spíritus: Trinity: A female, journey from Salto de Vega Grande to Salto del Caburní,

Topes de Collantes, December 6, 1991, a. Ávila C. (IES). La Sierpe: Una femma, Banao, October 30, 1992, unknown collector, no further data (IES).

Family Oxyopidae

Peucetia viridans (Hentz) (NRL). Cienfuegos: Cienfuegos: A male, La Legua, 13 km S de Soledad, August 1978, L. F. de Armas (IES-3.2333).

Family Selenopidae

Selenops aequalis Franganillo (NRL). Sancti Spíritus: Una hem-bra, Trinidad: El Chorrito, Topes de Collantes, July 11, 1975, R. Alayo (IES3.1646). *Selenopsaissus*(Walckenaer)(NRE). SanctiSpíritus :Trinity:Two females, road from Trinidad to Topes de Collantes, km 2,16 de December, 1973, L. F. de Armas (IES-3.1414 y 3.1415).

Selenops submaculosus Bryant (NRL). Sancti Spíritus: Trinidad: A female, 1 km E of Trinidad, August 1976, L. F. de Armas (IES-3.1239).

Family Tetragnathidae

Anelosimus jucundus (O. P.-Cambridge) (NRL). Sancti Spíritus: Trinidad: A female, Caburní stream, Topes de Collantes, August 1978, G. Alayón (IES-3.1587).

Leucauge regnyi (Simon) (NRL). Sancti Spíritus: Trinidad: three females, Caburní stream, Topes de Collantes, August 1978, G. Alayón (IES-3.1596 al 3.1598).

Family Theridiidae

Steatoda erigoniformis (O. P.-Cambridge) (NRL). Cienfuegos: Cumanayagua: A female, Boca de Carreras, August 1978, G. Alayón (IES- 3.2250).

Theridula gonygaster (Simon) (NRL). Cienfuegos: Cumanayagua: A female, Boca de Carreras, August 1978, G. Alayón (IES-3.1594).

APPENDIX II

Spider species that are endemic to Guamuhaya (*) or whose known distribution is restricted to this locality, but may be present in other areas of the country (**).

ARANEAE
THERAPHOSIDAE
*Cyrtopholis *bryantae* Rudloff

AMAUROBIIDAE
*Tugana *cudina* Alayón

CORINNIDAE
**Corinna Bryant, 1940
**Trachelas *contractus* Platnick & Shadab

CTENIDAE
*Ctenus *ramosi* Alayón

DICTYNIDAE
**Yorima *antillana* (Bryant)

DRYMUSIIDAE
*Drymusa *spectata* Alayón

DYSDERIDAE
*Gender undetermined, sp. n.

LINYPHIIDAE
**Gammonota *emertoni* Bryant

LIOCRANIDAE
**Phrurolithus *nemoralis* Bryant

PHOLCIDAE
**Anopsicus *pulcher* (Bryant)

**Modisimus concolor Bryant
**Modisimus *elongatus* Bryant

PISAURIDAE
*Dolomedes *Guamuhaya* Alayon

SALTICIDAE
**Agobardus *fimbriatus* Bryant
** *Agobardus prominens* Bryant
**Neon *nigriceps* Bryant
** *Thiodina peckhami* (Bryant)

THOMISIDAE
** Bryant's Tidy *Majellula*

AGORISTENIDAE
Piratrinus calcaratus Silhavy

BIANTIDAE
*Galibrotus *carlotanus* Silhavy

COSMETIDAE
*Trinimontius *darlingtoni* Silhavy

MINUIDAE
*Kimula *levii* Silhavy
*Minuides *milleri* Silhavy

SAMOIDAE
*Maracaynatum *cubanum* Silhavy
*Maracaynatum *stridulans* Silhavy

ZALMOXIDAE
*Cersa *kratochvili* Silhavy
*Ethobunus *cubensis* (Silhavy)
*Ethobunus *zebroide* (Silhavy)
*Pachylicus *castaneus* (Silhavy)

INSERT SEDIS

*Animota *custodiens* Silhavy

*Silhavy *longimanic* cariboule

*Silhavy *soft* valyphema

Printed by Books on Demand GmbH, Norderstedt / Germany